City of Sunderland Colleg

Hylton Learning Workshop

This book is due for return on or before the last date shown below
Please be aware that fines are charged for overdue items
Renew online: http://heritageonline.citysun.ac.uk
Renew by phone: call 5116231

checked.
13|5|2011 MM

GUIDES

B

Accession I + 68885   SPINING: Loc Code: CPD

Class 374.0120   Auth mcN

niace
promoting adult learning

National Institute of Adult Continuing Education
(England and Wales)
21 De Montfort Street
Leicester LE1 7GE

Company registration no. 2603322
Charity registration no. 1002775
Published by NIACE in association with NRDC

NIACE has a broad remit to promote lifelong learning opportunities for adults.
NIACE works to develop increased participation in education and training,
particularly for those who do not have easy access because of class, gender, age,
race, language and culture, learning difficulties or disabilities, or insufficient
financial resources.

For a full catalogue of all NIACE's publications visit
www.niace.org.uk/publications

Cataloguing in Publication Data
A CIP record for this title is available from the British Library
ISBN 978-1-86201-331-5

All photos posed by models, © istock.com

Cover design by Creative by Design Limited, Paisley
Designed and typeset by Creative by Design Limited, Paisley
Printed and bound by Latimer Trend

Developing adult teaching and learning: Practitioner guides

This practitioner guide arose from research and development work carried out by the National Research and Development Centre for Adult Literacy and Numeracy (NRDC) on young adults' learning of literacy, language and numeracy (LLN). It is designed to support practitioners working with young adults in a variety of settings with guidance chiefly drawn from the following NRDC projects:

- The role of informal education in developing literacy, language and numeracy skills for young adults
- Improving the literacy and numeracy of young offenders and disaffected young people
- Evaluation of the front-end delivery of basic skills in modern apprenticeships.

The guide also incorporates relevant findings and recommendations from NRDC's work on embedded learning and teacher education. For more information on NRDC's work in all these areas, please see **http://www.nrdc.orq.uk**

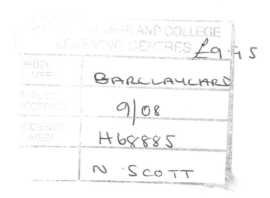

Contents

1 Introducing the guide

This book looks critically at how emerging and published NRDC research findings around language, literacy and numeracy work with young adults aged 16 to 25 can inform the development of teaching and learning strategies for this age group. It is designed to support practitioners working with young adults in a variety of settings. It will examine the main messages from NRDC research and development projects in this area and consider how they can be integrated into language, literacy and numeracy work with young adults.

The book's aim is to promote reflection, discussion and engagement with the research findings, in considering how they fit with day-to-day practice. Although it includes suggestions and ideas from the research, and practitioners' voices on how to integrate the key messages into practice, it is not a 'how-to' book. One of the strongest messages to arise from research consultation events with practitioners is that there is rarely a clear answer or catch-all solution in work with young adults. As practitioners at the events said, when it comes to what works with young adults, 'it depends'. It depends on the setting within which you are working, the learning programme being delivered, the aims and aspirations of the learners, and, above all, it depends on the individual learners themselves. As such, this book will not attempt to offer answers or instructions that are relevant across all contexts and settings. Instead, it builds in learners' and practitioners' 'stories', which will help to contextualise the messages from research, and the responses they triggered. These stories are fictional composites, but are based entirely on learners and practitioners who took part in NRDC research projects. The key messages will be filtered through these stories, to enable an exploration of the impact of these findings on different learners and practitioners in different contexts.

Background to the research

The NRDC's research in this area includes three major projects, which worked with a variety of practitioners[1] and learners, and across a variety of sites and institutional contexts. In addition, a range of NRDC research in other areas such as embedded language, literacy and numeracy and teacher education also has relevance to this area. There is increasing understanding of the importance of recognising the young adult age group, and of acknowledging and appreciating their values and cultures as distinct from other age groups. Similarly, increasing numbers of practitioners and organisations are working to support young adults in developing their language, literacy and numeracy skills, and this work is being integrated into more and more contexts where young adults can access support and IAG (Information, Advice and Guidance), vocational learning and training, and sporting and arts activities. Young adults are not resistant to learning itself, but are clear about how and what they want to learn. Practitioners who work with them are aware that motivation and engagement are critically important, and can be an ongoing challenge.

Young adults are involved in language, literacy and numeracy learning in a wide range of contexts and settings. For young adults on vocational programmes, literacy and numeracy, or key skills, are a core component of their learning. For those in custody, or serving sentences in the community, education and training, with an emphasis on literacy and numeracy, is a central strategy in reducing offending. Many colleges offer English for Speakers of Other Languages (ESOL) learning programmes aimed at young adults. Increasingly, voluntary and community-based organisations, and youth services, are recognising the potential of language, literacy and numeracy learning to contribute to social inclusion and active participation in society. The expansion of language, literacy and numeracy work with young adults has increased demand from practitioners for research and development work to support their practice.

Who are the learners?

Although NRDC research in this area has a common focus on the 16–25 age group, young adult learners are not a homogeneous group. They take part in learning in a variety of settings and contexts, and with a variety of aims and outcomes. They can also be referred to learning provision via an increasing number of routes and agencies. The learners' stories built into this book represent some of the most common experiences and backgrounds of young adult learners, in very different contexts. These learners' stories will provide a setting for the research findings throughout the guide, encouraging us to think critically about their translation into practice.

Who are the practitioners?

Possibly more than in any other area of teaching and learning, practitioners supporting young adults in developing language, literacy and numeracy come from very diverse occupational backgrounds and training routes, and take very different roles. This presents a number of challenges in sharing experiences and practice across the field, but also creates many opportunities for learning across sectors. Like young adult learners themselves, practitioners working with this age group are a far from homogeneous group, and as such, have varying training and support needs. The practitioners' stories that appear in the book provide some fictional examples of experiences and challenges, based on practitioners involved in NRDC research projects.

2 What does the research say?

An increasing number of practitioners are working with young adults to develop language, literacy and numeracy. New practitioners are training to become subject teachers and support workers, and practitioners who have worked with young adults for some time are seeking out professional development opportunities to enable them to support language, literacy and numeracy work with this age group. Other practitioners who have worked in this field for many years are undertaking continuing professional training or professional development, at all levels. Reflecting the growth in embedded language, literacy and numeracy provision for young adults, vocational teachers are increasingly taking on new knowledge about supporting the language, literacy and numeracy needs of their learners. This book aims to support practitioners working at all levels, and across settings, by identifying some of the main themes in work with young adults.

This section of the book sets out the key findings from NRDC research, and critically unpicks these messages, building in the comments from practitioners gathered at regional consultation forums. Each short section concludes on some practical questions to consider – how practitioners might integrate the suggestions into their own practice, and how the key messages might lead them to question or reflect on their practice.

The key messages are divided into three sections, focusing on the learners, the practitioners, and on teaching and learning approaches. Within these sections, each theme is explored, and situated within learners' and practitioners' stories.

The learners

Using age as a defining approach

What does this research say?

■ The term 'young adults' denotes an age group which can 'get lost' between young people and adults, and goes beyond the current focus on 14 to 19. 'Young adults' is used to refer to 16 to 25-year-olds.

■ It is challenging to find provision focusing on the learning needs of young adults aged 19 to 25.

■ It is critical to acknowledge the individual needs, hopes and experiences of young adults, as distinct from other age groups.

The 16 to 25 age group, it seems, can fall between two camps - on the one hand, we have seen a renewed concentration on the 14 to 19 age group in terms of education and training, particularly that which is employment-related. On the other hand, we also have a strong and sustained focus on adult literacy and numeracy. In practice, though, this often means older adults – employees, parents and the long-term unemployed, for example. At the same time, the literacy and numeracy needs of young adults are not always at the forefront of public consciousness – this age group tends to be conceptualised in other ways, and they occupy a contested terrain of rights, responsibility and governance, shifting between childhood and adulthood. While public concern about this group's educational experiences is often expressed implicitly, it is their behaviour which causes most anxiety. So, it can be seen that the 16 to 25s occupy a space between several arenas, and can consequently be overlooked, especially where research into language, literacy and numeracy is concerned. The term 'young adult' is used to distinguish between the younger age group (young people up to the age of 16) who fall under a different set of educational priorities and political drives, and to denote a specific age group within the broad post-16, adult education remit.

There was general support from practitioners at the consultation forums for widening the definition of 'young people', and a sense that age 19 is too fixed as an entry point to adulthood, and as a cut-off for provision. Many 19-year-olds are still unsure of what they want to do in life, and have not been involved

in education and training since leaving school. Practitioners report a big gap in services for this group for whom adult-oriented provision is not always appropriate.

There is also a growing belief that the term 'young adult' should include 14 to 16-year-olds, since FE colleges are increasingly working with this age group. It can no longer be assumed that 14 to 16s are involved in 'traditional' educational provision in schools, working towards their GCSEs. However, there are obviously strong differences between 14 to 19s and the 20-plus age group. This has prompted concerns that the term 'young adult' encompasses too wide an age range. Many practitioners feel that it is quite appropriate for the 20 to 25 age group to be included in the wider adult age group, since they can be facing very different priorities and challenges.

However, using age as a defining approach can be problematic, and practitioners stress the need to view all learners as individuals, concentrating on the methodology rather than the age in selecting the approach. As one practitioner commented, 'Young adults have different needs to adults, but it's dangerous to think that all young adults are the same.'

What does this mean for young adult learners?

Dean is 20 years old, and has been referred to a local training provider by Jobcentre Plus, through the New Deal for Young People. Growing up, Dean was sole carer for his mum, and consequently missed a lot of school. He feels that he was not supported to remain in school, and finally left at 15, before he had taken his GCSEs. He had short-term and low-paid jobs in the local market for three years, before signing on, and has been unemployed since. Dean lost his mum when he was 19. His girlfriend has just had a baby, and they have recently moved to their own flat. Dean is worried about supporting his family, and wants to 'learn a trade', which he feels will give him access to long-term and well-paid employment. He is working towards an NVQ in general construction operations, and enjoys the workshop sessions and hands-on nature of the learning programme, where he can see the process he is making week by week. He is confident that he can 'get by' in literacy and numeracy working in a construction job, and struggles to get motivated to improve these skills, becoming restless and resentful in literacy and numeracy sessions. His perception is that time spent on these areas is time away from the workshop, holding him back from achieving his NVQ, and employment goals. He loses interest quickly, and becomes disheartened and angry when faced with an exercise or activity he finds difficult. Dean is also preoccupied with his new family and their flat, which needs quite a bit of work to make habitable, and often asks for time off to sort these issues out. He rapidly forgets what he learned in previous literacy and numeracy sessions, and is rarely fully 'on task'. Without substantial work on his literacy and numeracy skills, Dean will struggle to pass his NVQ, although he does not see this.

At 20, Dean falls outside the 16 to 19 age group that is often thought of when the term 'young adult' is used. Dean sees himself very much as an adult, particularly now that he is a father. His caring role as he was growing up also meant that he made the transition to responsibility and adulthood relatively early in his life. He is very focused on his learning goals and participates fully in workshop sessions, getting irritated with the younger learners in the group who are easily distracted and misuse the equipment. However, Dean is still dealing with a range of challenging issues that can make it hard for him to engage with the learning programme. He needs ongoing support to resolve these issues, and relies on the training provider staff who have experience in assisting young adults to live independently, and make the transition into employment. Dean would not have remained in a learning programme without support, as his frequent absences would have caused too many problems.

Mark is 22 and is living in a hostel for young adults with substance misuse problems. He started using drugs when he was 14, and was excluded from school shortly afterwards. His mum asked him to leave the family home when he was 17, and he has been staying at friends' homes and various hostels ever since. He was recently referred to his current hostel where he also has a support worker, and feels he is finally ready to leave drugs behind. He has a bed in the hostel for 12 weeks and does not know where he will go at the end of this time, since he no longer has any contact with his family. His friends are still involved with drugs, and he is trying not to see them anymore, although it is hard because life at the hostel can be boring. There is an IT suite in the hostel, staffed by a tutor, and Mark can use this when he wants to play games on the PC or listen to music, in return for preparing a CV, and working through some entry level IT qualifications. Most of the time, Mark feels unmotivated and depressed by his situation, and finds it hard to see how working on a CV (when he has little work experience) can help him. If the tutor tries too hard to engage Mark in this work, he will walk out and refuse to return for some days. Mark does not like the idea of attending college, and gets defensive and withdraws when the tutor suggests referring him to a specialist literacy and numeracy provider.

Mark, at 22, has very little time for the younger adults at the hostel. When he decides to do something, he wants to get on with it without distractions. He sees the younger residents as immature time-wasters. He is not interested in attending learning programmes where there might be young adults, or where he might be classed as a young person. Mark had a difficult time as a teenager, and feels he is past this stage now, having experienced a great deal of life. However, Mark still does not feel ready to join adult classes at the local college, as there is too much going on in his life at present.

Anil is 16 and attending his local further education (FE) college. He is enrolled on an IT course, having previously tried a motor-vehicle maintenance course and not enjoyed it. He is reasonably interested in IT, but prefers to spend time talking to his friends about music and clothes. Although Anil sat and gained five higher-grade GCSEs, he has difficulty with literacy and numeracy, particularly in writing and spelling. He does not enjoy the written parts of his course, and needs a high level of support to write his essays and projects. He finds this boring, and is embarrassed at needing extra support. As a result, he can be uncooperative and abusive to the tutor and support lecturer when his friends are present. Anil has no clear ideas about his future career and dislikes discussing what he might do after the course has finished. At present, he is struggling to complete the work required on the course, and spends most of the sessions in conflict with the tutors.

At 16, Anil is very unsure about where he wants to go in life, and is happy living with his mum and dad, although he argues with them when they push him to work harder. Anil sees himself as free to do what he likes, with no responsibilities. He enjoys socialising, music and clubbing. The adults in his life pressure him, and hold him back, and he has no interest in spending time with anyone other than his peers.

Key issues for consideration and reflection

■ Is there a point when a young adult becomes an adult? Is this point different for every individual? If so, how can these differences be supported in learning programmes?

■ Is there a case for younger and older adults learning together to encourage learning from each other? Young adults may benefit from hearing the life experiences of learners like Mark. How can this be incorporated into learning programmes?

■ Do funding restrictions mean that young adults over 19 will always be amalgamated into the wider adult age group, regardless of whether or not this is appropriate?

Engaging and motivating young adults

What does this research say?

■ Engaging young adults is an enduring issue of paramount importance to practitioners. This is often more pressing than the language, literacy and numeracy elements of provision.

■ When considering publicity for learning programmes, young adults are more likely to believe and trust their friends than other forms of promotion.

■ Incentives are important for young adult learners. These incentives must be swift, achievable and tangible.

Most learning programmes for young adults aim to reach those not in employment, education or training, and this group, by their very nature, can be very difficult to find, let alone engage. Many practitioners report that finding ways to engage and motivate young adults must come before creating the language, literacy and numeracy elements of provision.

Developing credible publicity that attracts young adults can also be difficult. Young adults are more likely to engage positively with learning programmes, and remain engaged, when friends, family and boyfriends/girlfriends have also been involved. This seems to overcome any perceived stigma around their attendance, and makes young adults feel more confident and in control. Although young adults may initially attend provision when referred by intermediary or sign-posting organisations, research suggests that learners are more likely to drop out if the referral came from someone in authority, 'part of the system'. Learners are more likely to believe their peers, and feel more confident that their judgement is relevant to learners' lives, and made in their best interests.

Effective publicity is equally important when young adults are referred directly to provision (where the element of recruitment is less important), since it helps to develop the credibility that leads to sustained engagement.

However, balancing the initial 'hooks' and related rewards with the learning elements of provision can be a constant struggle. Young adults are a particularly hard group to engage, with their involvement often balancing on a knife-edge.

It is not only initial engagement that is critical, but also keeping learners engaged throughout the programme. Experienced practitioners tend to see incentives as essential, but they often add several key provisos. Incentives should:

- only be offered to engage learners in courses they are ready for – it is important that any later rewards are achievable (e.g. that a preparation for work course leads to employment);
- be built upon, so young adults are attending learning programmes because they want to; and
- be continuous, for example, through human contact and the relationship with the tutor.

Practitioners also acknowledge that the concentration on engagement can be a distraction, taking attention away from making the learning programme appropriate and accessible to all learners. Financial incentives can cause particular problems, in that they can become more important than the learning itself, leading young adults to 'pay lip service to learning'.

Practitioners agree that peer endorsement and word of mouth is much more effective than contrived 'youth' publicity. However, they also recognise the value of celebrity and role model endorsement, and highlight the importance of involving parents, where possible.

What does this mean for young adult learners?

Dean felt that the hook that pulled him into the vocational course he attends was the chance to gain a trade. The course was strongly marketed to him by Jobcentre Plus as being his escape from unemployment, and a means through which to support his family. He is motivated to attend, although sometimes everyday life intervenes, and he finds it hard to focus on the learning programme. Dean had not really considered that literacy and numeracy might be key parts of his programme, and is in danger of not completing his NVQ unless he makes significant progress in these areas. Dean is beginning to realise this, and is becoming increasingly resentful of this aspect of his course. Since Dean does not enjoy the literacy and numeracy part of the programme and does not always engage fully with activities, he may feel very disheartened if this means he cannot gain the NVQ he wants.

 **Anil** finds it very difficult to get motivated about his course. His tutors spend the majority of each lesson trying to engage him in the activities and the subject, but do not always succeed. Anil and his friends were allowed to listen to their music at the beginning of the course and, for a short time, this had a positive effect. However, the music soon became a distraction and caused disagreements within the group. The tutor withdrew permission to play music, and Anil's motivation is now worse than before since he perceives the tutor as going back on her word, and restricting his rights. Anil's incentive to attend sessions is receiving his Education Maintenance Allowance payments (which require 100 per cent attendance) and to see his friends.

Mark didn't return to the IT suite for a week after the first time that his tutor at the hostel mentioned 'basic skills' to him. Mark resented the intrusion into what he saw as his personal life, and was angry that the tutor was so quick to mention his poor literacy and numeracy skills. Mark still does not feel entirely comfortable in the IT suite, but he goes in to play games, surf the Internet and listen to music to ease the boredom of life at the hostel. Mark is considering what he wants to do to improve his chances of finding work and permanent accommodation, but he will talk to the tutor about this when he's ready, and when he's sure the tutor won't just send him to another provider for some unsuitable course. Mark does not spend much time with the other residents at the hostel, so is not really aware of the education and training activities they are involved in.

Key issues for consideration and reflection

■ How important is it to be upfront with young adults about the language, literacy and numeracy elements of their learning programmes? Is it a critical part of establishing trust or is it a turn-off?

- Can the use of incentives like computer games compromise time spent on learning activities? How can you strike a balance?

- What is the best way to establish the credibility of a learning programme where young adults' attendance is mandatory?

- Is it appropriate to involve parents in young adults' learning activities?

Young adults' perceptions of learning, and of language, literacy and numeracy work

What does this research say?

- Learning which appears to young adults to be irrelevant is often assumed to be boring, and results in swift loss of interest.

- Many young adult learners are likely to be wary of formal education, and feel that they know all they need to know.

- Young adults are not resistant to learning itself, but will reject provision they perceive as being too formal or school-like.

- Learners' primary interest is rarely language, literacy and/or numeracy.

The majority of young adult learners with language, literacy and numeracy needs have had a very negative experience of school and formal education. Most left compulsory education before 16, and sometimes missed extensive periods of schooling before leaving. Consequently, young adults are often reluctant to return to learning programmes they see as being like school, or to work with practitioners who remind them of teachers. Many feel they have 'finished with all that' and know all they need to know to get by in life as an adult and at work. Even at 16, young adults can already have a sense of failure, and this is a strong barrier to further engagement and achievement in learning.

Despite this, young adults are often keen to learn vocational and employment-related skills that will help them find and remain in work. Similarly, young adults are highly motivated to learn when the programme is linked to elements of their lives, experiences and interests. Events such as getting a job, moving into a first flat or becoming a parent provide the motivation, both intrinsic and extrinsic, to get involved in learning. Ensuring that learning programmes have a more holistic approach means that learners' interest is sustained through a concentration on their individual needs and interests, rather than on the potential end result or qualification.

Practitioners know that language, literacy and numeracy are essential to everything that we do. However, young adults do not always see this relevance, and it needs to be clearly demonstrated. When this does not happen, learners are quick to switch off. It is critical that practitioners themselves are clear about how language, literacy and numeracy relate to the learning programme, and the interests of the learners, and communicate this. However, whilst a focus on experiences or events such as preparing for work or parenthood can build motivation, it is also important that learning programmes take into account the realities of these experiences for young adults. Young adults need to be

motivated to remain involved in learning long term, and to see how it contributes to the development of their self-esteem and identity, rather than simply addressing immediate issues.

Bravado or a lack of confidence can lead young adults to assert that they know all they need to know, but this changes as they open up and grow in confidence. As they may only have experienced one 'negative' way of learning, it is critical to question and explore teaching and learning strategies, and present different styles and approaches to learning, to prove that there is life after GCSEs. Young adults, particularly the younger age group, often do not see themselves as learners within a formal context. However, practitioners report that young adults over 19 are far less resistant to formal learning, and can see an informal approach as a meandering route to achievement.

Young adults' primary motivation in attending learning programmes is rarely the development of language, literacy and numeracy, particularly where the basic skills elements of a course are stand-alone or discrete. They prefer integrated courses, particularly those with a vocational or creative slant. However, this is less true for young adults taking ESOL courses, where their primary and explicit aim is language learning.

What does this mean for young adult learners?

Leon is 18 and serving a 15-month sentence in a Young Offenders Institution for his involvement in a burglary. This is his third sentence. He is currently attending literacy, numeracy and life skills classes three afternoons a week in the education department. Although Leon would like to get a job when he is released, and does not want to return to prison, he does not enjoy attending classes, and repeatedly says he is bored. He gets distracted very easily and often refuses to actively participate. He did not attend school regularly after the age of 12, and when asked, cannot think of anything positive about his school experiences apart from meeting his friends, and occasionally playing football in PE lessons. He does not intend to take part in any learning programmes after his release. Leon has particular difficulty in understanding forms and written instructions, and refuses to engage with activities he thinks are pointless. He is very fatalistic, believing that he will never find employment on release because of his criminal record. This often overshadows any enthusiasm he may have for gaining qualifications, which he feels are an important part of finding work. He hangs around with four other young men in the prison, who are known for being very disruptive in classes. Leon hates the fact that he has no choice about attending classes, and feels that he is forced to learn things he doesn't want to. He occasionally takes part in enrichment activities, such as drama or music workshops, and comments that time seems to go quickly in these sessions, whereas it drags in other classes. Leon has a very strong sense of his identity, and resists personal change, seeing it as 'not being true to himself'.

Leon cannot think of anything that he learned at school that he has used since, and feels the same way about the education classes he attends at the Young Offenders Institution, where he feels that they do exercises 'for the sake of it'. He knows that qualifications are important for finding work, and is only interested in courses that will result in him gaining a recognised qualification as quickly as possible. However, he does not like the formal classroom environment or the lack of physical activity in his English and maths classes. He enjoys himself more, and time moves faster, when he is taking part in enrichment activities like drama, but he resents the fact that these courses are not accredited. He has commented to one of the tutors that he knows he is learning during the enrichment sessions, but 'it does not feel like it'.

Grace is 16 and attending an Entry to Employment (E2E) programme provided by her local youth service. She quite liked school, but was very badly bullied and felt she never fitted in. She consequently missed a lot of lessons and left as soon as she was able to. Grace was diagnosed with attention deficit hyperactivity disorder at the age of 11, and finds it extremely difficult to sit through sessions and to concentrate. She also acknowledges that she is easily wound up, and has got into trouble by physically threatening other learners when they provoke her. She is desperate to have a career in the RAF, and knows exactly what she needs to do to work towards this aim. She has strong self-belief, but realises that she can be her own worst enemy. She enjoys art and sport, and is very focused on improving her literacy and numeracy, although she prefers to work on these areas alone, away from the other learners. Grace does not know where she will go when her E2E programme finishes, as she is nervous about returning to a formal education environment, because of her experiences of bullying at school. Grace keeps herself to herself, but is articulate and open in one-to-one situations. She works much better with focused attention from a tutor and gets into difficult situations with other learners very quickly otherwise.

At school, Grace liked classes where she could 'have a go' and learn new things, like science and home economics. However, she struggled to learn in the traditional classroom environment, and found herself constantly in trouble for disrupting other students and for moving around. She is doing really well on her E2E course because she can learn more at her own pace, and has more freedom to move around the youth club. She wants to work towards qualifications to enable her to pursue her dream of a career in the RAF, but doesn't want to go to the local college because it reminds her of school, and she thinks the problems that she experienced there will recur.

Kelly is 17 and attends her local youth club most afternoons. She lives at home with her mum, and her boyfriend Sean has just moved in with them. Sean works in a shop and they share his wages with her mum to pay some rent and keep. Kelly hopes they will get married next year. She left school last year with two GCSEs, but doesn't feel confident enough to look for work, and is happy to be at home with her mum and Sean when she is not at the youth club. She likes going to the youth club in the afternoons, although she doesn't have many friends there since the other young women tend to pick on her. Kelly was told at school that she has a learning disability, which she thinks explains why she finds maths and English hard, but she doesn't know what it is. She is happy to be involved in the sessions at the youth club, which include II and media awareness, cookery, budgeting and art, but prefers to talk to the tutors rather than take part in the activities. Kelly has been attending the youth club for two years, and never wants to leave.

Kelly loves spending time with the group leaders and the few friends she has made at the youth club. Recently, the group leader suggested she might like to take part in a new course to help her with her English and maths, using computers. Kelly feels very nervous about this. She knows her English and maths aren't very good, but her boyfriend and mum help her with anything she needs to do. The other pupils laughed at her when she struggled at school, and she doesn't want to go through that again. Kelly has told the group leader that she's not interested in the English and maths course.

Key issues for reflection

■ How can we ensure the high degree of co-operation and team working required to demonstrate the relevance of language, literacy and numeracy across the curriculum, or where they are embedded within other subject areas?

■ Many young adults are not ready for independent learning. Does this necessarily mean adopting a more formal approach?

■ How can we demonstrate the knowledge and skills young adults need as adults without further damaging their confidence as learners?

Attitudes to accreditation

What does this research say?

■ Young adult learners are enthusiastic about gaining qualifications which they perceive have currency, particularly relating to employment.

The meaning and value of accreditation to young adult learners is highly individual, and often personal. Many young adults participate in learning programmes to move towards the type of employment they want, and obtaining certain certificates, awards and/or qualifications can form an important part of this strategy. However, the certificate, award or qualification offered must hold a value from the young adult's perspective, both in aiding the learner to move on (practically and ideologically), and in the eyes of potential employers. It can be challenging for practitioners to 'sell' certification, awards or qualifications that are believed by young adults to be of low or no currency. This is an ongoing difficulty, but can be overcome by providing a range of certification/qualification options, and emphasising their contribution to developing the 'whole person', including a range of skills highlighted by employers, such as negotiation and communication skills. It is also clear that for some young adult learners, certificates rewarding good attendance, team working, or personal improvement, do enhance confidence and promote a sense of pride. But it is critical for practitioners to know their learners in order to judge the value of such rewards. The most highly valued qualifications tend to be those directly related to practical activities that could be applied to obtaining employment. However, it is important to be aware that not all young adults will have the primary goal of moving towards employment: young mothers and fathers, for example, can be highly motivated in working towards certificates or awards that support them in their role as a parent.

Young adult learners are taking an increasing interest in the currency of qualifications, and are becoming more astute. They want to know what they are going to receive at the end of programmes, and how this will help them to move forward. However, many practitioners have found that issues of accreditation can also overpower the learning, causing learners' needs to 'get lost'.

The way that accreditation is introduced is critical, and for some young adults, it is important that they are fully engaged with the learning programme before accreditation or qualifications are presented. Acknowledgement of wider participation and achievement – perhaps through certificates - can also be important.

What does this mean for young adult learners?

Leon is very clear about what he wants from the education classes he attends at the Young Offenders Institution (YOI): qualifications that will make him more attractive to potential employers, and help to balance the negative effects of his criminal record. Although he enjoys taking part in activities like drama and music, he always asks the group leaders at the beginning of the course if he will gain accreditation through taking part, and is disappointed if the answer is no. He knows that literacy and numeracy are probably the most important subjects to get qualifications in since he didn't sit any GCSEs. He has received certificates for taking part in drama and music activities at the YOI, and although he enjoyed the activities, and feels quite proud of the certificates, he is cynical about how employers will view them.

Grace regularly receives certificates from her E2E programme, recognising her attendance, punctuality and improvement. The youth club holds presentation ceremonies several times a year. Grace doesn't like going up to the front of the group to receive the certificates, and plays down their meaning when asked. She keeps them at home in her bedroom, but would never show them to anyone as she doesn't think they have much value, particularly in relation to her aim of a career in the RAF. She is most interested in nationally recognised qualifications, and was proud of passing her Level 1 National Tests in literacy and numeracy.

Kelly feels very anxious when the group leaders at her youth club talk to her about qualifications. She didn't get any at school and feels that she is not clever enough to get any now. She feels under pressure when others in the group are working towards qualifications and hates sitting any kind of test. She was pleased when she received a certificate for 'most helpful group member' at the end of her cookery course, and took it home to show her mum. This was the first certificate she had ever received.

Key issues for reflection and consideration

■ When is the most appropriate moment to introduce issues of accreditation and qualifications to young adult learners? Is it possible, or desirable, to do this on an individual basis?

■ In learning programmes where qualifications are required for funding purposes, is there the flexibility to negotiate their introduction with learners?

■ How can links be developed with employers to develop their knowledge of the value of informal and non-formal awards for young adults?

The practitioners

The effects of isolation

What does this research say?

■ Many practitioners working to develop language, literacy and numeracy with young adults are outside colleges, and are often operating in isolation, with no access to professional networks.

■ Practitioners are keen to share their experiences, but there is not necessarily a common understanding of terminology used to talk about their practice (for example, embedded, non-embedded, informal, non-formal).[2]

The majority of practitioners working to support language, literacy and numeracy development with young adults are constantly searching for support, ideas, innovation and inspiration. Those who are new to the area, or who have relatively little access to training, are particularly eager to learn about best practice, or 'what works' in engaging and motivating young adult learners. Although many young adults are engaged in language, literacy and numeracy learning within FE colleges, there is an increasing amount of provision for this age group in other settings, such as youth clubs, foyers and hostels, prisons and probation, and within the voluntary and community sector. Practitioners in these settings are least likely to be in a professional network that links to their language, literacy and numeracy work with young adults.

Isolation can therefore be a real issue for those working outside FE, and there is a perception that networking and information exchange is much stronger within colleges. However, FE-based networks function most effectively for full-time practitioners, and even then, workload sometimes prevents full use of such networks. FE-based practitioners can therefore feel isolated too. One common complaint is: 'No one appreciates the role of the basic skills tutor... they think anyone can teach basic skills.' Management support to attend external events and networking is consequently important. The perceived separation between key skills and *Skills for Life* departments in some colleges is another factor that discourages networking.

Practitioners' networks can be useful for sharing knowledge, particularly locally, and in providing access to training and information. However, networks can easily become inaccessible because of geographical isolation or restrictions on membership (such as being qualified, for example). It is also true that networks are often most accessed by managers, and the information does not always get fed back to practitioners on the ground.

A lack of shared understanding and common definitions of practice can also impede networking. This is an important part of developing work and overcoming challenges. Confusion over terms such as 'embedded', 'contextualised' or 'integrated' can cloud issues for practitioners and learners alike.

What does this mean for practitioners?

Sam is a tutor in a Young Offenders Institution, teaching literacy and life skills. He finds his work challenging, but enjoys it, and takes great pleasure in engaging the young men in lively discussion about plays and books. He finds the environment challenging and restrictive, and feels that the training and education courses he is attended have not really prepared him for this situation. He gets little time off to train, and wants to know about training opportunities that will really help him in his role, but feels 'out of the loop'.

Sam's role means that he has regular access to training relating to security and other relevant issues, such as anger management. However, he has less access to training in approaches to teaching and learning, and isn't sure where to look. His Internet access at the prison isn't reliable, and he has to use his own computer at home to search for information and developments relevant to his work. Sam feels that external training he has managed to attend often does not seem to take account of the realities and restrictions of working in the custodial setting.

Patrick is an IT facilitator at a voluntary organisation working with homeless young adults. He recently left university with an ICT-related degree and is very confident in demonstrating a range of ICT tools and packages to the young adults. This has helped him build strong relationships with them. However, he had not realised how much support they would need with English and maths. Patrick doesn't see himself as a teacher, and didn't want to go down this route, but is now wondering how to develop his awareness and skills in this area. He has no links with the local college and is the only person in his project taking on this type of work. He doesn't know where to start looking for the right course, and is worried that it will take years of part-time study to gain the qualifications he needs, which may in turn make it harder to keep his ICT qualifications up to date.

Patrick has no idea about how to access professional networks to find out more about language, literacy and numeracy work with young adults. His manager at the hostel is very supportive, and is happy for him to attend any events that are appropriate, but Patrick does not know where to look. He has tried reading around the subject on the Internet, but finds the terms confusing, and is always looking for a straightforward guide to the key issues and terminology. He knows there must be others in a similar position locally, and nationally, but does not know how to get in touch with them.

Karina is a tutor at an FE college, working with young adults with ESOL needs. She is passionate about her work and has attended various training courses for ESOL specialists. She is the only person at her college who works with this group, and she is struggling to build links with other departments, to help them understand the support the learners need in their other curriculum areas. She feels that none of the training she has had has really addressed the issues involved in working with young adults, and feels isolated in her role.

Based in a college, Karina has access to professional networks, and is generally up to date on recent developments, research and new initiatives. However, she is most keen to forge stronger links with her colleagues in other departments of the college, and to share knowledge of her area and theirs, to create a more supportive environment for the learners. She is finding this difficult because of the heavy workload, and the separation between the departments within the college infrastructure.

Key issues for reflection and consideration:

■ What is the most effective way of overcoming feelings of isolation in FE colleges, encouraging different departments and curriculum areas to work together?

■ Is there a role for a national network linking practitioners supporting language, literacy and numeracy development with young adults, or is the need more locally based?

■ How can information be fed effectively to practitioners at all levels?

■ What is the role of research in developing a common understanding of language and terminology?

Tailored training and professional development

What does this research say?

■ Many practitioners working with young adults have very little specific training in the teaching of language, literacy and numeracy.

■ Practitioners in this area have a wide range of training and qualifications, and may not see themselves primarily as language, literacy and/or numeracy teachers/tutors.

■ Practitioners' roles are highly complex, and do not always fit easily into the roles within the Skills for Life Teaching and Qualifications Framework.

■ Personal qualities and attributes, such as patience and empathy, are considered essential for practioners; whereas language, literacy and numeracy training is seen as desirable but too often hard to access, and sometimes inappropriately tailored to the context.

- ■ Practitioners report that training they have undertaken has not always developed the skills and abilities they feel are most important in their work.

- ■ Practitioner training should build on existing skills and experiences. This training also needs to reflect and respect the variety of roles practitioners play. Such training is not widely available at present.

Training and professional development has emerged from NRDC research as an area where opinions across sectors are most likely to differ. Access to training and professional development, particularly that which is related to Skills for Life, varies widely across sectors.[3] It can be challenging for practitioners working outside FE colleges to gain up-to-date, accurate and useful information related to training. Even within colleges, research showed that many practitioners were working towards/had achieved generic teaching qualifications without appropriate language, literacy and numeracy qualifications, or were working towards/had achieved a language, literacy and numeracy qualification without generic teaching qualifications.

The pre-2007 Skills for Life teaching qualifications framework did not easily fit the context of practitioners in community or youth work settings, since roles are far from straightforward, and cannot easily be segregated into supporting the learner or the teaching process, or leading the learning. Many practitioners without specific teaching qualifications are leading the language, literacy and numeracy learning in their organisation with young adults. For these practitioners, the emphasis is on engagement and building relationships. However, practitioners working in this area are certainly not under-qualified in general, and bring vast amounts of varied and relevant knowledge to their roles. Practitioners in all sectors have completed a range of courses, qualifications and certificates, all of which support their work with young adult learners. For many practitioners working in the youth work and voluntary and community sectors, it is the case that language, literacy and numeracy-specific qualifications are not always considered the most important qualifications to seek for their role.

Practitioners working in the community feel strongly that the right people to work with young adults are practitioners initially trained as youth workers, whether or not they have any specific language, literacy or numeracy related training. Some also contend that the abilities and empathy required as a youth worker are not only paramount but somehow innate – qualities that individuals are born with and cannot learn through training.

Practitioners from the voluntary and community sector often find it hard to access training, either because of their heavy workload or a lack of funding. Consequently, some suggested that language, literacy and numeracy work with young adults is an area lacking organisation and control. Despite this, the richness in experience and skills is recognised, in part due to the complexity of roles in this area. Accessing language, literacy and numeracy qualifications and training was also seen as an effective way for practitioners working across sectors to come together and learn.

Practitioners regard training as important, partly because it helps to raise the status of their role, but also in providing invaluable confidence and support in relation to their work. However, it also has to be acknowledged that training and professional development can only go so far in preparing practitioners for work with young adults, for 'real life teaching'.

Practitioners feel that 'incredible skills' are needed in work with young adults, and they are right: this is an extremely difficult area in which to teach. A practitioner can be 'qualified up to the hilt and still have problems'. This may imply that something extra needs to be added to professional development activity to prepare practitioners to work with this age group.

What does this mean for practitioners?

Pat has been a youth worker for 15 years, working at the same youth club in a deprived area in the north-east of England. She loves her job, and is confident in working with groups of young people, and in creating exciting activities for them to get involved in. Recently, her youth club was awarded funding to develop literacy and numeracy programmes, with support from the local college. Pat has been asked to manage the project. She is very nervous about this, and feels that her own literacy and numeracy skills might not be up to scratch. As part of the funding, Pat can access learning programmes for herself at the college. She feels under some pressure to do so but does not see herself returning to education in this way. She likes working with the tutor from the college who visits her youth club, but also knows that in order for the programme to be a success, she must 'sell' it to the young adults and embed it into the other activities. At present, it is perceived as a 'bolt on' session, and is not well attended. She sometimes struggles to support the young adults when the tutor is not present.

Pat has been considering taking up the opportunity to work towards literacy and numeracy teaching qualifications as part of the project she manages, but has some concerns about her own skills. She is worried that she will be the only person on the course working in a youth club, and that it won't mean very much to her. Pat qualified as a youth worker many years ago and doesn't see herself as a teacher or tutor. She would really like to do some training aimed at youth workers that would develop her knowledge around Skills for Life.

Marie is a tutor at a large FE college and was recently asked to go out to a nearby foyer providing accommodation for homeless young adults. The foyer wants to provide a 15-week literacy and numeracy course for the residents. Marie has many years' experience in literacy and numeracy teaching with adults, and is highly qualified. As such, she was confident about developing the 15-week course, and excited, although a little anxious, about working with a new group of learners. She has now completed two sessions at the foyer, and feels she cannot cope. Marie had not expected the group to be so hostile, or so difficult to manage. She is not satisfied that the group are making progress, and she has had to abandon much of her course plan. Consequently, she has asked her manager at the college if she can leave the foyer, and communications with the foyer manager have broken down. Marie is concerned about how to respond if she is asked to take on such a role in future.

Marie is currently working towards her Level 4 literacy qualifications at the large FE college where she teaches adults. Marie is finding her work with young adults at the foyer extremely stressful, and her confidence has been knocked by the experience. She does not feel she is cut out for work with young adults, and does not think that any training would make her feel more comfortable in that role.

Carl is an ex-professional footballer, now working as a tutor and mentor at a private training provider. He really enjoys working with the young people, and they respect him, especially the young men. He did some work with local school-children at his former football club, and decided to move into this area of work when an injury forced him to retire from professional sport. Carl has no teaching qualifications but is happy supporting the young people in their learning. His manager wants him to start by enrolling on the City and Guilds 7407 at the local college, but Carl is not sure he will cope with the course. He knows he will have to address the issue of training if he wants to continue in the job.

Carl loves working for a training provider and has an excellent rapport with young adults. He has no teaching qualifications, and did not enjoy school himself. He feels this is one of the reasons he is able to empathise with the learners so well. He is not against working towards teaching qualifications, but does not have much confidence in his own abilities, and feels it is more important to continue developing the relationships he has with the young adults. Privately, Carl's manager also has concerns that the 7407 course may be too challenging for Carl.

Key issues for reflection and consideration:

- If 'something extra' needs to be added to professional development for practitioners working with young adults, what might this look like, and how can it complement Skills for Life and other qualifications, such as those for Youth Work?

- What professional development should be on offer to those whose primary role is not teaching Skills for Life, but who are involved in language, literacy and numeracy work as part of their role?

- It is widely agreed that personal qualities such as empathy and understanding are central in work with young adults, but can these qualities be developed through training?

'Not being like a teacher': The importance of building relationships with young adults

What does this research say?

■ Young adults need to feel that they can relate to tutors or leaders, and for most young adult learners, this means 'not being like teachers'.

The vast majority of young adults involved in NRDC research had overwhelmingly negative memories of formal education – of failure, bullying, authority, exclusion, mistrust and boredom. When asked what they enjoyed, or appreciated, about the subsequent language, literacy and numeracy provision they were involved in as a young adult, the most frequent response was 'it's not like school... X is not like a teacher'.

As has already been said in this book, effective engagement with young adult learners must take priority in order to sustain longer-term involvement with the learning programme. This involves creating the right environment and employing the right staff. The quality of relationship between practitioners and learners is of central importance, and practitioners working with young adults need to be 'user-friendly': aware of the types of issues young adults may be facing, and approachable and non-judgmental in their advice and support.

There is an acknowledgement that school did not work for many young adults, and consequently a different approach is necessary, with clear boundaries set in managing behaviour and raising expectations, but accompanied by holistic support for pastoral issues.

Some learners, however, are more comfortable in formal settings, and it is important to find the balance between challenge, discipline and security in learning, with clear boundaries and frameworks.

What does this mean for practitioners?

Karina finds that working with a range of young adult learners in an FE environment means adopting a variety of roles and approaches. She finds that some young adults, particularly on ESOL courses, arrive at the college looking to her as a strong group leader and teacher, and enjoy the formal environment of the class. Others, however, do not respond to this approach and she has to spend much more time on their pastoral support. Their relationship with her depends upon her creating an environment that is very different from the one they experienced at school.

Pat feels that one of the main reasons she has such a good rapport with the young adults she works with is that the youth club is a space where they can learn and explore issues, but is totally separate from and free of associations with school. The young adults who go to the youth club often talk to Pat about school, and the difficulties they experienced there. Pat feels pleased that she is able to offer them the support they need, and that they are happy to keep working with her.

Collette manages a learning zone attached to a Connexions shop in a small town. She is a qualified Connexions personal adviser and leads learning programmes for young adults excluded from school. She does a lot of literacy and numeracy work with the group. Her employer is supporting her to work towards a City and Guilds 7407 qualification, but she knows it is just the first step. She is worried that she does not have the time to commit to gaining qualifications, but is also concerned that she does not have enough experience at present to effectively build literacy and numeracy into the learning programme. She is constantly searching for new ideas and initiatives to engage the young adults, but doesn't really know where to start looking.

Collette knows that the learning zone has to be a very different environment from school, and that she must be something other than the traditional idea of a 'teacher'. However, Collette also knows that to maintain control of the group, and to make progress during learning sessions, she must ensure that there are clear boundaries in place, and that her role as group leader is explicit.

Key questions for reflection and consideration:

■ Is it possible to adopt an approach of not being like a teacher whilst maintaining discipline and security in learning?

■ Does the approach depend on the age group?

■ What are the implications for training in adopting this approach?

Teaching and learning approaches

Teaching by stealth or making language, literacy and numeracy explicit

What does this research say?

■ There is an ongoing debate among practitioners about the benefits of making language, literacy and numeracy explicit in learning programmes rather than 'teaching by stealth'.

■ If the practitioner dislikes or is negative about language, literacy and numeracy, young adult learners will be too. Treating language, literacy and particularly numeracy as something to be endured alongside the more exciting elements of the learning programme is a common, but negative, approach.

■ It is important to provide a clear rationale for the learning at the outset, one that makes sense to learners.

The manner in which programmes are introduced to young adults has strong implications for the language, literacy and numeracy elements of the learning. This has particular relevance on three levels: *whether* language, literacy and numeracy are referred to as part of the learning programme, *how* they are referred to, and how this is *rationalised* to learners.

In NRDC research, practitioners were divided over whether language, literacy and numeracy should be introduced to learners 'up front', at the beginning of the learning programme, or whether it should be disguised within other areas or subjects. Many were concerned that making the language, literacy and numeracy elements of the learning programme explicit too early (or at all) would switch learners off, causing low uptake and poor retention. For others, it was crucial to be upfront, and recognise the varying reasons why learners may be attending provision: for some, developing language, literacy and numeracy may be an important factor in achieving their learning aims and aspirations. Practitioners at the NRDC consultation forums favoured a middle-ground approach. As one said:

> Teaching by stealth is an odd feeling – it's introducing [language, literacy and numeracy] by the back door, but it doesn't have to mean secrecy... it just doesn't have to mean explicit literacy and numeracy either.

Language, literacy and numeracy should be made explicit at some stage during the learning programme, but this does not need to be at the beginning, nor does it need to be emphasised. 'Hiding it', or 'lying', is recognised by many as a negative and counter-productive approach, which often results in damaging relationships with learners. Being open and honest usually pays dividends. However, an element of surprise is often necessary. Reviewing progress made during the session and reflecting on what has been learned can be an effective way of introducing the language, literacy and numeracy elements of a programme without overtly focusing on them. This was seen as key in maintaining motivation. There was also concern at the forums that 'teaching by stealth' could become confused with an embedded approach, which is more explicit.

Although practitioners are divided over teaching by stealth, there is widespread agreement over the importance of providing a clear rationale for the learning at the outset of any programme. Stating the learning outcomes for every session is also now seen as good practice, as is keeping learners informed and involved in discussions about aims and objectives as they progress through learning programmes. Central to this practice is the positive projection of language, literacy and numeracy – where practitioners are negative about these areas, learners will be too. This has staff development implications: non-specialists need support in demonstrating the relevance and importance of language, literacy and numeracy to young adult learners. Language, literacy and numeracy must be shown to be personally useful, linked to young adults' goals, and applicable to their own lives. Fundamentally, learner need should direct the learning programme.

What does this mean for young adult learners and practitioners?

Dean, when he was referred to a local training provider by Jobcentre Plus to work towards his NVQ in general construction operations, didn't know that literacy and numeracy would be part of his learning programme. When he found out during the first week, he felt quite demotivated since he finds these areas difficult, and he was set on the more practical elements of his vocational course. He probably still would have come to the training provider if he had known, as he is very committed to completing the NVQ, but he feels a bit annoyed that it was not explained beforehand.

Grace knew, from her first day at the E2E programme, that she would be working towards the National Tests in literacy and numeracy. This was very important for her, since she knew that these qualifications would help her to find work with the RAF. Grace also likes to know in advance what it is she will be doing as part of her learning programme, so she can decide how it will be helpful to her. She does not like surprises, or being required to do something without being told why.

Pat is having difficulty in deciding how to introduce the literacy and numeracy courses to the young adults who come to her youth club. Having worked with young adults over many years, she is not sure that they will see the relevance or importance of the sessions, and may find them boring in comparison with the other club activities. She does not feel that confident herself in her own literacy and numeracy, and thinks she will have problems talking about these skills positively, and demonstrating the links to the young adults' lives.

Key questions for reflection and consideration:

■ How far is reflection the answer to the search for middle ground between teaching by stealth and making language, literacy and numeracy explicit at the beginning of a learning programme?

■ Is there enough flexibility within learning programmes to allow the needs of individual learners to shape the introduction of language, literacy and numeracy?

■ How can practitioners be supported to project language, literacy and numeracy positively?

The relative roles of language, literacy and numeracy

What does this research say?

■ Provision tends to concentrate mostly on literacy, with far less emphasis on numeracy. Oracy is often overlooked.

NRDC research has found that literacy tends to appear in learning programmes for young adults much more explicitly than numeracy. Numeracy is far more likely to be embedded into vocational subject areas, or specific life skills sessions, featuring most prominently as an aspect of financial awareness or budgeting. Practitioners seem to feel far less confident about numeracy, both in terms of their own skills and in the anticipated response from young adult learners. Consequently, numeracy is far more often an implicit element of sessions – contained within the activities but not readily apparent or drawn out. Although many practitioners say they also work to develop young adults' communication skills, the development of speaking and listening receives substantially less attention than literacy and numeracy. Although oracy development often occurs through many different strands of work, it is not regularly mentioned as an explicit, or conscious, element of provision.

Practitioners at the forums widely agreed that, outside ESOL programmes, oracy is often overlooked. For some, this is because of time or funding constraints: oracy is not assessed as part of the National Tests. For others, oracy is an integral part of learning programmes, but is not always focused on. As one contributor to a consultation forum said:

> Oracy is taken for granted because we're speaking and listening all the time. We rely on language so much, we can forget to teach them [speaking and listening]. We automatically assume learners are picking them up.

This is regrettable because oracy development is essential in work with young adults, particularly where the learning programme focuses on employability.

However, we should also recognise that many practitioners do not know the best strategies for teaching oracy. Some are also wary about it because it is such a personal skill. They fear that teaching specific ways of communicating could be removing learners' last bit of identity.

What does this mean for young adult learners and practitioners?

Kelly tends not to pay too much attention to the literacy and numeracy elements of the sessions she attends at her local youth club preferring to talk to the leaders or her small group of friends. Kelly knows she finds it difficult to learn and remember things, and doesn't think she'll pass any tests or certificates, so has decided not to try. However, she believes that her confidence has really improved since she started going to the youth club regularly, and she now feels she can have conversations with more people, and explain her feelings more clearly. She realised this when the group leader pointed it out during one of her review sessions.

Karina is accustomed to a focus on language, as she is working with groups of young adult learners as part of ESOL learning programmes, but she is also careful to highlight the speaking and listening elements of all the work she does, including literacy and numeracy. Karina knows that developing speaking and listening in the widest sense has a strong impact on the confidence and interpersonal skills of the group, and they notice this too, in their day-to-day lives.

Collette would like to do more work on speaking and listening at the Connexions learning zone, and tries to encourage the group to discuss and debate topics that interest them. However, she's not sure how to evidence this work, or to measure progress. She knows it is happening as part of the learning, but would like more guidance on how to develop it.

Key questions for reflection and consideration:

- How can we overcome some of the cultural issues surrounding numeracy (for example, 'it's ok to say you're poor at maths') to incorporate it as a core component of learning programmes?

- What research and development work is needed to support the integration and recording of oracy work into practice?

- How much can work on speaking and listening compromise learners' sense of identity? How can oracy work be developed, maintaining respect for identity and peer-group interaction?

The role of learning styles in work with young adults

What does this research say?

- Learning style questionnaires, used in conjunction with other such tools, can be effective in developing learning programmes for young adults.

- Learning should be practical and 'out of their seat', with authentic activities.

NRDC research has demonstrated the importance in work with many young adult learners of creating an atmosphere or environment that is not like school. Young adults who had negative school experiences often found the structure and style of learning problematic, and respond best to active sessions such as PE or drama. Similarly, NRDC research has shown how young adults respond positively to workshop or practical sessions as part of vocational programmes. Practitioners have also found that practical and authentic activities and opportunities for learners to be 'out of their seat' are important in helping maintain motivation.

Many practitioners agree that it is important to explore different learning styles, particularly in meeting individual learner need. However, learning styles questionnaires can be a crude tool in assessing individual styles and preferences, and can lead to fixed teaching strategies which do not reflect the needs of whole groups. Similarly, such assessments do not always reflect the fluid and changing nature of learning styles. Practitioners generally discover this it is more effective to establish a range of activities during learning sessions, through observation and ongoing negotiation with learners.

Practical, 'out of their seat' learning can be very effective but this approach should not dominate. Learning should be fun and involving, but it is important to recognise that this sometimes means learning being 'in seat'. In any case, funding and environmental constraints can make an 'out of their seat' approach impractical. Fundamentally, success hinges on learners recognising that learning is relevant, and a variety of approaches and styles may be necessary to build this ideology.

What does this mean for young adult learners and practitioners?

Mark is most focused on learning that he sees has close relevance to his life. Although he sometimes finds it hard to get motivated, he knows that improving his ICT skills at the hostel where he's staying will help him secure employment in the future. He doesn't like being in sessions where others are disruptive, and just wants to get on, working slowly through what he needs to do without distraction. He sees breaking up sessions with lots of activities as childish, and wasting time by not really moving him forward.

Leon feels that the thing he dislikes most about his education classes at the YOI is that he has to sit behind a desk for the whole session. He gets easily bored, and loses concentration. When he takes part in sessions that are more active time passes much more quickly and he realises he's quite enjoying himself. Leon knows he really has to focus to remember things when he's reading them, or listening to his teachers, but is able to make links much more easily when he's physically active and involved.

Patrick, in preparation for his work with homeless young adults, read a lot about learning styles and is very interested in their impact. He works hard to incorporate a range of learning styles and preferences in the sessions and tasters he runs in the IT suite. Young adults can be involved in activities like digital photography in the local neighbourhood, or spend time on the Internet researching areas of interest. He feels he strikes a balance between different types of activities, and that this is reflected in the young adults' high levels of engagement.

Key questions for reflection and consideration:

- ■ How can differentiated teaching and learning methods be used to support a range of learning styles in environments that are not conducive to 'out of their seat' learning?

- ■ How far is practical or experiential learning a 'meandering route to achievement'?

- ■ How can learning styles be most effectively explored without categorising learners?

Exploring embedded approaches

What does this research say?

■ Highlighting how language, literacy and numeracy relate to vocational areas, and using interests as a vehicle for embedding can have a strong effect on the perception that the learning programme is real, relevant and 'adult'.

■ Embedding or contextualising language, literacy and numeracy in other subject areas enhances learners' motivation, engagement and learning outcomes.

NRDC research has clearly shown that embedding or contextualising language, literacy and numeracy enhances learners' motivation and engagement. Using these approaches can build enthusiasm, interest and commitment, and may result in higher achievement and retention rates.[4] Language, literacy and numeracy are commonly embedded in sessions such as 'shop and cook', gym and fitness and sexual health. Embedding, or contextualising, language, literacy and numeracy within vocational areas is also increasingly widespread as more providers develop their 'vocational offer' in line with policy and funding priorities.

For most practitioners, embedding or contextualising language, literacy and numeracy is about making links for learners, demonstrating relevance and transferability. However, as previously noted, an embedded approach can be confused with 'teaching by stealth', and result from a drive to hide or disguise the language, literacy and numeracy elements. Some practitioners do this to reduce stress for learners, others do it because they lack confidence in their own language, literacy and numeracy. An embedded approach can also translate into an opportunistic approach, where the language, literacy and numeracy elements are identified retrospectively, and are unplanned or even unrecognised.

An embedded or contextualised approach can be very effective in work with young adults, but it must be done properly.

For example, it is critical that practitioners feel confident in the language, literacy and numeracy elements of their teaching, and are able to address problems in these areas should they arise during sessions. This has clear training implications. The potential unintended consequences of embedding have also been considered. Learners can become increasingly resistant to discrete provision and perceive it as childish in comparison with embedded or contextualised provision. There is also the danger that the language, literacy and numeracy elements can become so embedded that learners come away without the skills they need or without seeing the relevance of these skills. As one practitioner, who took part in the NRDC consultation noted, 'Improving literacy is often about raising awareness – you can't do this if it's too embedded.'

A discrete approach can also work, and indeed is sometimes more appropriate, particularly where learners' motivation rests on the achievement of a literacy, language and/or numeracy qualification.

What does this mean for young adult learners and practitioners?

Dean attends discrete literacy and numeracy classes as part of his learning programme at the training provider. He sometimes resents going to these sessions, and finds them challenging, but realises that he is much more interested when the tutor gets him to think about how the literacy and numeracy are part of the vocational skills he is learning. He finds it easier to solve problems when they are contextualised, and knows why he is doing it.

Grace is very focused on working towards qualifications as part of her E2E programme, and whilst she enjoys sessions on cookery and digital photography as she can participate and be active, she prefers to concentrate on her literacy and numeracy in separate, more focused sessions as she believes this moves her towards her goal more quickly. She knows she is learning things in the other sessions, but she finds it harder to see the relevance when it's not the explicit focus.

Collette, when she started running the learning programme at the Connexions learning zone, incorporated two sessions of discrete literacy and numeracy into the timetable. However, she found these sessions were the least successful with the young adults, who quickly became bored and questioned why they were being forced to do 'pointless work'. She has since begun embedding literacy and numeracy within their other sessions, such as sexual health, budgeting and life skills. Collette is not always sure that the learners know they are working on their literacy and numeracy, and feels this work sometimes gets lost within the sessions. She doesn't want to revert to discrete sessions but isn't sure how to effectively draw out the literacy and numeracy elements of the learning.

Key questions for reflection and consideration:

- Has the focus on embedded learning meant that discrete approaches are being overlooked, even where they are most appropriate?
- Is an embedded approach necessarily a more adult approach?
- Is there a need for training in embedded and contextualised approaches, outside existing qualifications and courses?

Formality, informality and workplace learning

What does this research say?

- Traditional classroom settings can impede learning.
- Learners respond well to programmes that offer perceived flexibility similar to the workplace, for example, making drinks and refreshments, taking breaks and listening to music.
- An informal approach is highly effective with young adults. Informality is about approaches to setting, curriculum, relationships, pedagogy and assessment. In practice, creating and sustaining an informal approach can often involve high levels of formal planning and structuring for practitioners.

Practitioners taking part in NRDC research have frequently raised the importance of the learning environment in work with young adults. The majority of practitioners seek to promote an informal learning setting, or at least the appearance of this, recognising that there can be high levels of formality behind the scenes. The main driver behind this approach is the desire to create an environment that is not like school. Playing music, having comfortable chairs, and ensuring the room is warm and welcoming can all help to achieve the right atmosphere. Aspects of an informal learning environment, such as being able to 'make a brew', have a chat or listen to the radio, are seen as characteristic of work rather than school, and are often as important to young adults. However, they also need to be clear about the realities of the workplace, which can be formal and restrictive.

An informal approach can also be seen as less appropriate when it comes to identifying frameworks for achievement, establishing boundaries and parameters, making sure learners are clear on who is managing and leading, working in a classroom environment with appropriate equipment, and maintaining structure and consistency rather than flexibility. Informality is seen as desirable and likely to encourage active learning when it fits in with formal programme aims, and does not make too many demands on time in preparing resources. Informal approaches can be useful in engaging young adults who are reluctant to get involved in learning programmes. As the work progresses, the approach can become more formal. A formal approach is often associated with a more adult approach and learners who are more focused on what they want to achieve. It can also help to create a feeling of security for new-arrival and ESOL learners.

However, despite this concentration on the classroom environment it is also important to continue focusing on effective teaching and learning strategies, as practitioners commented:

> If the style of teaching is effective, students will learn whatever the setting.
>
> Teaching is what makes the difference, not necessarily the environment.
>
> It's how the space is used that matters.

What does this mean for young adult learners and practitioners?

Anil believes he is treated more like an adult at the FE college he attends. He likes the fact that there is a common room, a canteen, and that he can smoke outside during his breaks. He still dislikes the more formal setting in the classrooms, and resents the fact that the tutor has stopped them listening to music in sessions.

Collette has worked hard to create a welcoming and appealing environment in the Connexions learning zone, and involved the young adults in choosing easy chairs, plants, posters and paint colours. She knows that establishing this environment was critical in the learners making the distinction between school and the learning zone. The learners are also able to make tea, coffee and toast in their breaks, and listen to the radio. However, Collette maintains a more formal approach during learning sessions, since she has found that structure and boundaries are very important in keeping the learners 'on task'.

Pat is clear that the informal approach adopted by the youth club where she works is very popular with the young adults, and that incorporating elements of formality into the setting would have a negative effect on their engagement with activities. Pat has a very relaxed attitude to her work, and takes pleasure in watching the young adults enjoy their surroundings and the sessions at the club, but she also knows how much time with her colleagues goes into planning activities, and the paperwork that accompanies them.

Key questions for consideration and reflection:

- Does an informal approach have to mean a lack of structure or boundaries in the learning programme?

- Is a formal approach the only way to make learners feel secure in the learning environment?

- Is research and development work needed to clarify terminology around informal, non-formal and formal learning?

3 Conclusions and implications for practice

This book has demonstrated some of the complexity in language, literacy and numeracy work with young adults, highlighting the importance of situating messages from research within the various contexts in which learning takes place. Practitioners at the consultation forums were very clear that this book should avoid reductive guidance and should not produce a set of instructions. Instead it should focus on the processes in work with young adults, exploring and thinking about the issues. There are no straightforward answers, or catch-all solutions in this work, and an effective approach will always begin with putting the learner's needs at the centre of developing provision and approaches.

Practitioners also emphasised that there is a wide range of exciting, innovative and promising practice in work with young adults, and that this should be recognised. This is central in developing the confidence of practitioners, learning about the strengths in the field, and in beginning to build the necessary networks. Training and professional development issues are raised throughout this book, but the message is that any new training must build on the good practice that already exists, complementing current training packages and balancing established professional development routes.

Practitioners were asked what messages they would like to see emerging most strongly from this book. The overriding message was clear – young adults are a diverse cohort with diverse needs, and this, along with variations in setting, context and professional background of practitioners, must be recognised. Effective approaches in work with young adults start with this recognition, and build up provision around it, incorporating reflection, negotiation, respect and creativity.

Notes

[1] Throughout this publication, 'practitioner' is used to cover a range of different roles in work with young adults, including teacher, tutor, trainer, youth worker and mentor. This is explored further in the section 'Who are the practitioners?'

[2] For further exploration of these issues, see Casey, H. *et al.* (2006) *'You Wouldn't Expect a Maths Teacher to Teach Plastering...' Embedding Language, Literacy and Numeracy in Post-16 Vocational Programmes – The Impact on Learning and Achievement.* London: NRDC.

[3] For further discussion of these issues, see Lucas, N. *et al.* (forthcoming) *Towards a Professional Workforce: Adult Literacy, ESOL and Numeracy Teacher Education 2003–5.* London: NRDC.

[4] See Note [2] above.

Further reading and sources of information

Aylward, N. (2005) *Lessons from the Adult and Community Learning Fund.* Leicester: The National Youth Agency.

Casey, H., Cara, O., Eldred, J., Grief, S., Hodge, R., Ivanič, R., Jupp, T., Lopez, D. and McNeil, B. (2006) 'You wouldn't expect a maths teacher to teach plastering...', *Embedding language, literacy and numeracy in Post-16 Vocational Programmes: The Impact on Learning and Achievement.* London: NRDC.

Cranmer, S. and Kersh, N., with Evans, K., Jupp, T., Casey, H. and Sagan, O. (2004) *Putting Good Practice into Practice: Literacy, Numeracy and Key Skills Within Apprenticeships. An Evaluation of the LSDA Development Project.* London: NRDC.

Hurry, J., Brazier, L., Snapes, K. and Wilson, A. (2005) *Improving the Literacy and Numeracy of Disaffected Young People in Custody and in the Community.* London: NRDC.

Jackson, C. (2003) *Working with Young Adults.* NIACE: Leicester.

LSC Skills and Education Network Guide 2...
- *Engaging Young Learners: Entry to Employment (E2E) – Good practice*
- *Engaging Young Learners: Entry to Employment (E2E) – Policy and strategy*
- *Engaging Young Learners: Improving young learners' achievement*
- *Engaging Young People: Overcoming the barriers*
- *Engaging Young People: Policy and strategy*
- *Engaging Young People: The Educational Maintenance Allowance*
- *Engaging Young People: What motivates young learners to engage*
- *Engaging Young People: Young mothers*

Available to download at **http://senet.lsc.gov.uk/guide2/Guide2.cfm**

Lucas, N. *et al.* (forthcoming) *Towards a Professional Workforce: Adult Literacy, ESOL and Numeracy Teacher Education 2003–5.* London: NRDC.

Marken, M. with Taylor, S. (2004) *Back on Track: Successful Learning Provision for Disaffected Young People. Good Practice Guidelines.* London: LSDA.

McNeil, B. and Dixon, L. (2005) *Success Factors in Informal Learning: Young Adults' Experiences of language, literacy and numeracy.* London: NRDC.

Websites

Basic Skills Agency at NIACE: **http://www.basic-skills.co.uk**

Maths4Life: **http://www.maths4life.org**

National Research and Development Centre for Adult Literacy and Numeracy: **http://www.nrdc.org.uk**

National Literacy Trust: **http://www.literacytrust.org.uk**

National Youth Agency: **http://www.nya.org.uk**

Read Write Plus, Department for Children, Schools and Families: **http://www.dfes.gov.uk/readwriteplus**

Young Adults Learning Partnership: **http://www.niace.org.uk/research/yalp**